POWER TOOLS

Thoughts about
power & control
in service to people with
developmental disabilities

Dave Hingsburger

> *Consistency requires you to be as ignorant today as you were a year ago.*
>
> Bernard Berenson

Diverse City Press Inc.

Diverse City Press (La Presse Divers Cité)
33 des Floralies
Eastman, Quebec   J0E 1P0

450.297.3080   www.diverse-city.com

© Copyright 2000
By La Presse Divers Cité
All rights reserved

Hingsburger, Dave

Power Tools: Thoughts about power & control in service to people with developmental disabilities

1. Developmental Disabilities
2. Power Dynamics
3. Staff Training

ISBN 1-896230-18-0

> *Why doesn't everybody leave everybody else the hell alone?*
>
> Jimmy Durante

Table of Contents

Introduction: A Dark and Stormy Night	1
Thinking About Power	4
Myth-Taken Assumptions (1)	5
Myth-Taken Assumptions (2)	7
Myth-Taken Assumptions (3)	11
The Process of Change	15
Step One: Only You	16
Power Play (1)	19
Step Two: Another Think Coming	21
Power Play (2)	23
Step Three: Detection through Introspection	27
Power Play (3)	29

> *Who Know?*
> *Maybe my life belongs to God.*
> *Maybe it belongs to me.*
> *But I do know one thing:*
> *I'm damned if it belongs*
> *to the government.*
>
> Arthur Hoppe

Skill Number One: Scene Not Heard   30

Skill Number Two: Power Sharing   35

> *A man is either free or not. There cannot be any apprenticeship for freedom.*
>
> Imamu Amiri Baraka

Introduction: A Dark & Stormy Night

It got dark, fast. Real fast. We found candles. Bought a flashlight. Lit a fire in the fireplace. The phone was still working so we called Hydro. A nice young man answered, he was without a care in the world – our situation didn't change the colour on his mood ring. As the sun set we opened a couple tins of beans and placed them by the fire. Beans are a humble meal, but heating with natural gas is sometimes necessary.

Shortly after, we were comfortably watching the fire's dance cast shadows on the wall. The phone rang. We chatted with a friend and complained of our plight. We couldn't do anything that we wanted to do. E-mail was out of the question. Our daily fix of *Law and Order* was put on hold. Heck we couldn't even read the newspaper. Just before hanging up the caller said, "It must be awful to be without power." I quickly agreed.

The fire was mesmerizing. The sounds of Hurricane Floyd's winds whipping against the trees made the darkness, just outside the fires reach, feel like a blanket wrapped round us. We were protected here. I thought of the day and remembered something. It was such a common sight that I didn't even really notice. But now the image was strong in my mind. A little boy being pulled along by his mother as they walked outside the grocery store in Montreal. He stopped and pointed to something.

> *You shall have joy, or you shall have power, said God: You shall not have both.*
>
> Ralph Waldo Emerson

She yanked him so hard that you could almost feel the muscles of his arm scream in pain. Then they were gone.

It must be awful to be without power.

Eric, the youngest dog, crawled up on the couch with me and turned around three times. Relieved that he didn't find vermin in the folds of the blanket that covered my lap, he lay down. My hand found his ear and in seconds he was snoring softly beside me. Again, from the small crack in the corner of my mind, another image came to me. "Who do you think you are!?!" Her voice was so hard. Her tone communicated that the only answer she wanted was, "No-one, I am no-one." The old woman in the wheelchair could have been her mother, the woman pushing the chair was so young to have such a bitter voice. I don't know what sin the old woman had committed, but being old was sin enough. She looked down and watched time pass along with the cracks in the pavement.

It must be awful to be without power.

The beans were hot and bubbly and we fetched plates, some bread was buttered, and we settled down to eat the meal. It wasn't what we would have made, but it was good. Better than expected. Eric sat up and watched us eat, waiting for the bits of bread that he knew would be his treat. I turned and stretched out on the couch. Eric lay beside me on the floor. And again, I remembered. The group home looked nice. In fact it looked way too nice. Everything was neat and in its place. Only a government home looked like that. He, however, wasn't government issue. His clothes fit. Sort of.

> *The greater the power the more dangerous the abuse.*
>
> Edmund Burke

His walk worked. Sort of. His life mattered. Sort of. He had come into the kitchen to get something from the fridge. Just as he opened the door, a young male staff looked up from his paperwork in the office and yelled, "Get away from there? Can't you see I'm busy here?" Forgetting for a moment that this man's need formed the basis of his job, the young staff slammed the fridge door shut. "You know the rules. No snacking, get back to the front room." There was no protest, just shuffled steps moving away from a bit of cake in the fridge.

It must be awful to be without power.

The lights came back on at 11:30. They wavered a couple of times over the next few days, but they stayed on. Power. It's an interesting thing. It must be awful to be without it, true. But what of those who misuse it? What of those who have it, horde it and hurt with it? It must be awful to be without power. True. But it must be lonely to live without conscience.

> *Powerlessness frustrates; absolute powerlessness frustrates absolutely. Absolute frustration is a dangerous emotion to run a world with.*
>
> Russel Baker

> *I am more and more convinced that humans are dangerous creatures and that power, whether vested in many or a few, is ever grasping and like the grave, cries, "Give, give."*
>
> Abigail Adams

Thinking About Power

First let's recognize something. Even though I wrote the above piece for my regular column in my local newspaper with the intent of showing how commonplace the abuse of power was, I was kind of fudging. I don't really believe that most people realize that ...

A) they have power
B) they routinely abuse that power
C) their behaviour is invisible only to themselves
D) their responsibility isn't diminished because they "didn't mean to ..."

A black woman activist, her name lost to history, writing in the United States over a century ago said, "All power seems natural to those who hold it." This is an astounding observation. This means that once you have power in your grasp it seems so natural, so normal, so "the way it should be" that you don't even realize it's there. This means, that I don't think that the mother who yanked her child's arm truly understood what she was doing and who she was harming. Similarly, though the woman in the wheelchair was clearly demeaned – I don't think that was the intended result of the interchange.

Finally, with the staff in the group home, I don't think he's a mean person. In fact, I know him, he's a cool guy who really cares about the people in his care. But in the moment, the moment that he was distracted and interrupted, he reacted. And when he reacted he forgot who he was, what he was there to do and how his words can affect someone so dependant on his good will.

> *Power always thinks it has a great soul and vast views beyond the comprehension of the weak; and that it is doing God's service, when it is violating all Divine laws.*
>
> John Adams

Ignorance of power doesn't imply that there is no responsibility for what you do. This little book is aimed at getting you to think about what you do on a daily basis with those within your care. This is particularly important for those who work at the front line of human need and in the service of human care. Those who do this work are often the most powerful person in the lives of those they serve but the most powerless person in the helping organization for which they work. (I really think that the term "helping organization" is a contradiction in terms, humans help, organizations organize – when we get that, we might want to take another look at the salary grids!!)

Myth-Taken Assumptions (1)

Even so, feeling powerless makes it more likely that you will make powerful mistakes. You may develop a *personal mythology about personal power*. You might say to

> *Our sense of power is more vivid when we break a spirit than when we win a heart.*
> — Eric Hoffer

yourself, "but I don't have any power. I have lots of responsibility but I don't have power." I know this from experience. LuAnne was a woman with a disability who I worked with in a pre-vocational centre (that was before Lou Brown had taught us that "pre- means never") who was "over-social" which in behavoural terminology for "likes to talk as much as the staff do." We decided to put her on a behavioural programme and what with me fresh out of university and fresh into a new job, I jumped at the chance of writing the programme. (The irony of me writing a programme for someone who talks too much escaped me at that moment, but *I had no time for introspection, I was busy fixing someone else.*) I knew from the pages of a text book that time-out was an effective punisher.

Well, we didn't have a time out room so we adapted one of the programme rooms to become the LuAnne isolation room. Then whenever she was "over social" we locked her in that room and demanded 5 minutes of silence before she'd be let back out into the "common area" (which lacked "common decency," "common courtesy," and "common sense"). And it worked. And I got a promotion. My first. Yep, I was rewarded for systematically silencing a woman with a disability.

Not once, in my work with LuAnne did I stop and consider who she was, who I was, the dynamics of our relationship and why it was that I had a right to change her to meet my standards when it would never be possible for her to change me to meet hers. That imbalance alone should have had me quaking in my boots as I picked up a pen to determine both goal and plan. But it just seemed like the right thing to do. I had more power than I thought. My personal mythology of power had exempted me from considering myself as a possible power-monger, because I saw power in terms of the hierarchy above me, not in terms of the paper upon which I wrote, the pen in my hand which outlined the programme in my head, and the procedures that we would be allowed to do to someone, for her own good of course.

> *Power takes as ingratitude the writhing of its victims.*
>
> Rabindranath Tagore

Myth-Taken Assumptions (2)

Knowing I don't fit into airplanes with the greatest of ease, I always wait just before it's boarding time and approach the friendly (?) service (???) agents at the gate. I ask them to check my seat assignment and see if there is someone seated next to me. If there is, please move me. If there isn't, praise be. The thin, thin gate attendant, had lips pursed so tightly that she looked like she'd rather have sex with a camel than go down on a Popsicle said, "Yes you are seated next to someone." I smiled benignly, I am very tolerant of those whose rib cages echo empty and asked, "Could you see if you could find me a place where there wouldn't be someone next to me?"

> *Resistance to tyrants is obedience to God.*
>
> Thomas Jefferson

She clicked a few more buttons and asked in a tone hard enough to chip a diamond, "You want a whole row?" Well, I don't need a whole row, but I'll take a whole row so I said, "Why, yes, thank you so much" and boarded the plane. I found my row and popped myself in the seat by the window. The armrest was lifted and I was comfortable. Just then a tall black man boarded the plane in a real rush. He is told to take any available seat. He swung himself comfortably into the aisle seat of my row.

Now you have to understand. I'm a guy. I had peed on the four corners of my row and now he was sitting in MY ROW. He said something to me and I was frosty to him. *HE WAS IN MY ROW*. We took off into the air and I relaxed into the ride. Suddenly I realized that he must think I'm a racist. After all, I'm fat. I'm white. I gave attitude. But I'm also Canadian, we were the *end* of the Underground Railroad. I pondered how to bring up the fact that I'm not a racist, but gave up anything I could think of would sound stupid.

Then we hit bumpy air. I had something to talk about. "Bumpy isn't it?" says I with a charming grin. He was frosty back to me. I deserved that, but I'd started and that's all I needed. I quickly explained and he smiled. "Yeah, transit is kind of like that," he said adding, "have you ever been on a bus and when people are coming towards you your mind says over and over, 'Don't sit next to me! Don't sit next to me!' then when they pass you think 'What's wrong with me? What's wrong with me?'" We laughed and talked. He told me that he was a student priest. Cool. In bumpy air you want to be sitting next to a student priest. As we talked, he discovered that I worked with people who have disabilities. He fell silent for a moment and then said, "Do you want to see a picture of the face of God?" Now you are in an airplane, sitting next to a priest, offering you a picture of the face of God, who's gonna say no? He paused saying, "When you are a victim of prejudice yourself, it's difficult to go public with your own prejudice." Then he pulled out the picture and offered it to me.

> *In order to obtain and hold power, one must love it.*
>
> Leo Tolstoy

My worst fears were realized. God looked like a forty-fiveish pasty faced accountant. There really is a ledger book, I thought, and someone's counting up my sins. Yikes.

The priest then told me of teaching a class about God to a group of men and women with developmental disabilities at a large institution. At first, he said, he was so overwhelmed by their "difference" he stumbled. When he had taught the same class to "others who weren't *others*" he always began by having them close their eyes and imagine an all powerful and all seeing being. Then the discussion usually focussed on how most see God as white, as old, as male. Can people begin to see God as like themselves, can God be seen with a black face? with a woman's hands?

> *The love of liberty is the love of others, the love of power is the love of ourselves.*
>
> William Hazlett

These are profound questions and he loved asking them and watching people struggle to answer. But when the first group of came through into the institution classroom he couldn't imagine asking them to ponder similar thoughts. However, by the third or fourth class he had begun to see holiness in both individuals in the group and in the group itself. So he asked. He asked each member of the group to close their eyes and imagine an all powerful person. When they opened their eyes he asked them to describe this person. One man began to describe in some detail what 'God' looked like. It became clear he was describing someone he knew. In response to the question, "Who is this man?" the man with the disability walked the man with the collar down the hallway and pointed at the social worker's office.

"I carry this picture of the man who worked in that office with me at all times. I need to remind myself that *we who say we serve have tremendous power over those we say we serve.*"

I pondered that all the way down to the baggage claim area and when we were getting our luggage I said to my seat-mate, "You know I'm glad you were in my row." He looked at me and said, "No, I'm glad *you* were in *my* row."

Oops.

We have power over those we say we serve. It's funny isn't it. We say we are in it for those in our care. We say we everything we do we do it for them. We use the word "serve"

> *Power? It's like a dead sea fruit; when you achieve it, there's nothing there.*
>
> Harold MacMillan

like a waiter would use the word. A waiter knows his job, his obligations and for whom he works. But when we use the word "serve" I don't think we often think of people with disabilities as our employers, we think of them as our raw ore, the skills we teach as our product and the behaviours we suppress as quality control. Their need gives us our power. Their vulnerability, our control. Their disability, our reason.

Myth-Taken Assumptions (3)

I ran into LuAnne ten years later. Coming home from the beer store on a Friday night, I looked up the street and there she was. Walking up Wellesley Street from the subway with a friend. She was talking a mile a minute and her hands flapping in the breeze her mouth made. I couldn't believe it. She looked just the same.

When she got close enough to hear me, I said, "Hi, LuAnne how are you?" She glanced at me and I saw recognition in her eyes. A quick answer flew out of her all in one word, "HiI'mfinethankyouverymuchforasking." Then. Then. Then. She walked away from me. I had spoken to her and she walked away from me. My mouth opened and these words came out, "LUANNE, get back

> *Power will intoxicate the best hearts, as wine the strongest heads. No one is wise enough, good enough, to be trusted with unlimited power.*
>
> Charles Caleb Colton

here." She stopped, as much at the tone as the words, I'm sure. Turning she walked back to me, silenced. "LuAnne, I asked you how you were." Without looking at me she said, "I said I was fine."

"You know LuAnne, you haven't changed a bit." And she hadn't, she looked exactly has she had ten years before. Then she said it, and I tell myself she really didn't mean it – I do have to sleep at night – she said, "You haven't changed either."

I haven't seen her since.

I had come to believe that old myth number three, you outgrow the temptation to misuse power. Like, come on, I'd by then taken courses in "Feminist Approaches to Psychology" and had studied power dynamics. Like, I'd gotten an A, eh? Like, I wouldn't use a tone of voice that would make King Tut feel a belittled. Like, I'm so *over* power and control. Like, **Not**.

Well, coming to believe that I'd conquered in me the worst part of the human condition was nice while it lasted. Hey, we just love to give give give ... there's nothing in all this serve serve serve for us. Hey, look at my paycheck, I don't have control. Hey, look at my humility, I don't need control. Hey, look at my devotion to society's under trodden, I don't want control. ***I SAID LOOK.***

> *It is by the promise of a sense of power that evil often attracts the weak.*
>
> Eric Hoffer

I learned that when I look at my profile in the mirror, my resemblance to a pudgy Mother Theresa is obvious only to me. And what with that halo, I don't need a nightlight – during daylight hours.

However, I do learn. The incident with LuAnne really bothered me. I began doing shifts in group homes and sheltered industries and keeping a kind of diary. I would keep a record of things that I said whenever my power or control was tested. I marked down everything that came out of my mouth when patience ended and words began. First, the words. Here are some things I actually said, in little brackets below I've written what I actually meant ... we all know that we say one thing and mean another.

Well it's about time!
        (... stupid!)

Because I said so!
        (I'm in control here.)

I said "Right Now!!"
        (There is no room for debate.)

I don't have time for your nonsense!
        (There is no room for you here.)

> *Tell me what you think you are and I will tell you what you are not.*
>
> Henri Frédéric Amiel

Then when I was taking courses in counselling, the professor made all of us tape our counselling sessions so that we could listen and evaluate our work. The idea of taking time to evaluate my own work seemed wasteful! Why would I need to do this? Didn't she realize that I worked with society's downtrodden and didn't have power and control issues? Well, YIKES! When I listened to the first tape there was no way, no way in hell that I would take it to class.

And it wasn't cause I said anything stupid, after all I knew I was being taped. It was because of *how* I said what I said. After all the years of training, all the inservices I'd attended on communication, all the books I'd read about power dynamics, I didn't expect to hear what I heard. There on the tape I was challenged about something and something odd, almost *X-files-ish* happened. Margaret Hingsburger, my mother, came out of the tape recorder. That voice that my mother used that let me know that if I didn't stop right now, she'd have "my guts for garters," (she actually said that!) came out of the tape recorder. How my mother got in there, I don't know. What I learned though was that *I was invisible to me.* That I had to watch extra hard all of my interactions with people with disabilities to ensure that I was doing what I thought I was doing. I also learned that this was possible.

I wish I knew where LuAnne was these days so I could tell her that I had finally changed!

The process of change.

> *Every one thinks God is on their side. The rich and powerful know he is.*
>
> Jean Anouilh

There are really only three steps, and two skills, that lead to change. I know that you've all been led to believe that there are twelve steps, but I'm fat – if it's over three steps, I take the elevator. The trouble is that this all takes a great deal of effort and determination. I've always believed that power was perhaps the most addictive drug. Especially when it's injected directly into the vain. So expect to go through some withdrawal.

A personal note to you from Dave's friends, family and dogs: Please, please, please warn those you love and care about that you are about to go through this little bit of personal change. You may end up working so hard to control your control at work that you come home and try to get your daily dose of power in your interactions with us. Not Fair. We can help you grow too you know!! Now we aren't saying that Dave likes to control much more than the television remote – of course not. We just wanted to add our voices in here at this particular moment.

> *The only way to predict the future is to have power to shape the future. Those in possession of absolute power can not only prophesy and make their prophecies come true, but they can also lie and make their lies come true.*
>
> Eric Hoffer

Before we begin, I think it's important to say something quite serious. You may think that because I'm writing this booklet, I'm also saying that power and control issues no longer affect me and that I'm so self aware that I don't have to worry about this sort of thing any more. Nothing can be further from the truth. I am very wary of Myth-Taken Assumption Number 3. I will never be done with this. I find myself slipping all the time. In the last few days (none of this is false humility, it's just the truth) I've misused my power with a clerk in a store, a member of my family, a hotel staff; I've done it in conversations and with actions; I've done it willingly, knowing full well what I was up to; I've done it and only come aware of it hours afterwards. Those people who are "over" power are really only "over" honest self-evaluation. So this is me now. In fact, step number one comes from something that hit me in the face the day before yesterday.

Step One: Only You

Toby is a cool guy. A half-day's growth on his face, worn jeans, and a quirky grin. He's hell to teach. He'd come with his girlfriend to a workshop that I was giving for couples with disabilities. He cracked jokes all the way through and, though it was clear he loved his girlfriend –

he teased her almost non-stop. The first thing he said when introducing her to me was, "You know I was normal until I met her!" He was so open about his disability that it was somewhat unnerving.

> *The dispensing of injustice is always in the right hands.*
>
> Laurie Lec

After the workshop was over I asked Toby if he enjoyed himself. He said that he'd had a good time. Then he looked at me, almost like he was taking a measure of my ability to take criticism. Then wham, "Only thing is retarded people don't do things the way you teach." I was shocked. First he used "that" word without even a hint of blush or embarrassment. Second, he was suggesting that "retarded people had a distinct way of doing things that I didn't get." Third, how dare he.

"What!?!" I said with some temper probably entering my voice, after all he had been a distraction for much of the class.

"Yeah, you want us to use too many words. Me, I ask my girlfriend if she wants to fuck. If she says "yeah" we do, if she says "nah" we don't. It's easier than what you teach. There isn't all this ... all this ..." he was lost for words.

"Process?" I suggested.

"Yeah. You should teach this again but next time teach it the way we do it, not the way you do it."

17

> *Power is so apt to be insolent, and liberty to be saucy, that they are seldom upon good terms.*
>
> The Marquis of Halifax

It seems that Toby figures that I did OK in teaching, but what I'd taught them was how I did things when they needed to learn how to do things in their own time and their own way. I get this little slap in the face pretty much once or twice a year, but Toby put it in such a way that I could finally understand it. I figure that everyone is pretty much like me (or should be) and therefore they just need to do what I do in order to get by. My whole teaching curriculums come from the view point of a normal though fat white guy. My measuring stick for normal comes not from a great survey of what all people do, but from understanding myself as a representative of the "norm." I hadn't figured either that I could be a bit "off centre" myself, or that people with disabilities might have their own norms.

When I was making the movie, "No! How!!" with a bunch of self advocates from the Essex County Association for Community Living in Ontario, it became clear to me that people with disabilities knew what they needed to learn and that they did indeed do things slightly differently than I would have done them. And the movie is a hit because the authentic voice of people with disabilities came through.

So what's the step in simple terms? Recognize that your job isn't to craft people with disabilities into your own image. This will lead to frustration and to battles of will and power. You end up with the use of force. Simply recognizing that the only thing that's important is that people with disabilities discover their own way in the world, you can let go of a lot of frustration.

> *You only have power over people so long as you don't take everything away from them. But when you've robbed a man of everything he's no longer in your power – he's free again.*
>
> Alexander Solzhenitsyn

Power Play (1)

The Scene: A kitchen of a group home.

The Action: A man with a disability has just finished doing the dishes. As the staff enters, the man has just folded the dishcloth and hung it over the sink tap. The dishes rest on the dish rack.

The Players:

Robert: 46ish year old man with Williams Syndrome

Martha: 18 year old-ish woman with rings on her fingers and rings in her nose.

The dialogue:

Martha: What do you think you are doing?

Robert: *confused* Huh?

Martha: *annoyed* I asked, What do you think you're doing?

> *Every man likes the smell of his own farts.*
>
> Icelandic Proverb

Robert: *still confused & just guessing* The dishes?

Martha: *getting angry* Don't get smart with me. Look at that counter top and tell me what's wrong.

Robert: *looking at the counter* I'm finished the dishes.

Martha: *patient and exasperated* Come on Robert, just take a look. You've got to learn how to do this right.

Robert: *desperate* I wiped down the counter.

Martha: *patronizing* Robert everyone knows that you stack the large plates at the back of the rack. You've got the saucers mixed in with the plates.

Robert: *giving up* I'll never get it.

Martha: *encouraging* You've just got to try harder, Robert.

First thing you have to ask yourself is, why does Martha care how the plates are placed in the rack? Does it really matter. Here, Robert had done all this by himself and instead of feeling good (or even thanked) he ends up feeling like a hopeless failure.

Second thing you need to ask is, what should Martha have done differently? (Other than get rid of the nose ring – at first glance nose rings always look like snot that has become encrusted on someone's nostril, but maybe that's just me.) I figure that Martha shoulda just noticed that the dishes were done. So what if they aren't in a nice neat row? So what if the task analysis says that dishes should be in a nice neat row? So what if every normal person in the world would put their dishes in a nice neat row? So what? The dishes are done. The counter is wiped. The dishrag is well hung. So what about the way the dishes are stacked?

> *I have never been able to conceive how any rational being could propose happiness to himself from the exercise of power over others.*
>
> Thomas Jefferson

"Problem is, Dave, retarded people don't do it that way." I'm never going to forget that. I know that the statement will offend the politically correct police, but it's an amazing statement. He did the dishes. That's all that matters. Martha trying to make Robert into a model of herself is just an abuse of her power. And he ended up victimized by her expectations rather than celebrated by her approval. A real tragedy.

Step Two: Another Think Coming

I was facilitating a group of self advocates to write a bill of rights for their agency. We were gathered in a large room about 40 self advocates, a front line staff who acted as a secretary and myself. I have done a few of these by now and have found that most of the time self advocates want similar things. Privacy. Relationships.

> *Self advocacy is discovering the 'our' of power.*
>
> Dave Hingsburger

Decision Making Power. That kind of stuff. However, one man struggled to put into words what he really wanted from the agency. Finally he had it and he blurted it out. "I want the staff to take the extra five minutes it takes for us to figure things out." I didn't realize that I'd be blown away by this statement until hours later. How damned perceptive!!

People with disabilities in that room knew that they were being short-changed. They knew that the staff were taking the time they needed to write programmes, go to meetings and to neatly compile incident reports. But when it came time to actually let a person with a disability work through something it's "*hurryhurryhurryhurry* don't you realize I'm *busybusybusybusy*!"

Five extra minutes. That's what he asked for – to a solid round of applause from everyone else there in that room. Now what would he want the time for? I'm guessing it's not to hear us talk, or give directions, or offer an opinion. I'm guessing it's so that he, and others, have time to process information and think for themselves. Have you ever had someone push you to respond when you are still trying to sort out your thoughts in your own mind? If you have I'll bet you've said, "Hold on a second, I'm still thinking." It takes time to think. Now remember it may take a little longer when a person has a cognitive disability (that's what a cognitive disability is by the way). So, they have another think coming, and they want to wait until it gets to the station.

Problem is, when we run out of time, we usually just supply an answer. We don't even notice that we HATE this when someone does it to us. Ever notice that when you are sittin' and bitchin' about someone, the last thing you want is advice or to be told what to do? Well when someone comes home from work and they are upset and want to talk about what happened at work. Why not just listen, let them finish, let them come up with a solution? I know that it's easier to say, "You need to tell your supervisor that Philip is bugging you," but why say it?

> *The wrong sort of people are always in power because they would not be in power if they weren't the wrong sort of people.*
>
> Jon Wynne Tyson

And moreover what happens when you say it?

Power Play (2)

The scene: Grocery Store Aisle

The action: A woman with a disability has just picked up a bag of cookies and is placing it into the grocery cart.

The players:

Susan: a tubby woman who walks with a slightly odd gait

Hank: a tubby man who walks with a slightly uptight gait

Janet: a fellow shopper not known to Hank or Susan

> *It's not difficult to explain to white mice why black cats are unlucky.*
>
> Graffiti

The dialogue:

Hank: *loudly* Susan put those cookies back.

Susan: *hesitantly* But ...

Hank: *quickly and loudly* Susan, come on now we don't have time for this sort of thing. We've got to get back to the group home.

Susan: *quietly but firmly* But ...

Hank: *interrupting* There are no buts about it. Put the cookies back.

Susan: *quickly trying to get a word in* But I like these ...

Hank: *judgmentally* You know the doctor said that you were getting fat. I said put the cookies back.

Susan: *giving up* But ...

Hank: *firmly knowing he's won* Right now.

Susan: *sadly* Sigh (and puts the cookies back)

Janet: *while passing by asks* Is it really that bad for her to have some cookies?

Hank: *explaining* She has to learn to make new food choices.

Janet: *walks away saying* Looks like you need to learn the same thing.

Hank: *referring to Janet under his breath* Bitch

Susan: *smiles*

> *It's not wise for man to make widowhood a woman's only route to power.*
>
> Gloria Steinhem

The first thing you need to ask yourself about this situation is, "*What?*" There are so many things here that it's difficult to know where to begin. So let's make a list:

A. Why would he speak so loudly and in a public place about something as private as weight?

B. Why wouldn't he let her finish her sentence? (My friend who has cerebral palsy says, "Having a disability means never getting to finish a ..."

C. Why would he use authority rather than reason as a way of dealing with the situation?

D. Why didn't he anticipate the issue and have all this worked out before hand? Did they have to walk down the cookie aisle?

E. Why, as a fellow porker himself, didn't he realize just how difficult putting cookies back on the shelf would be for her?

F. Why did he see that Janet was doing to him exactly

25

> *With friends like this who needs enemas?*
>
> Graffiti

what he was doing to Sue?

But the problem really is much deeper than all these things put together. What say Sue really did need to start making different decisions. His way of "helping" really isn't help, it's control. Wouldn't it have been best for her to figure out what should go in the basket and what should stay on the shelf? How much longer would it have taken him to ask her to stop and think before the purchase? Does she really want the cookies? If she does, is she worried about what the doctor said? If she is, what kinds of things could she eat that would give in to her sweet tooth and still be OK? (And don't do the carrot sticks and an apple kind of thing – OK? **OK?**) If she wanted to buy the cookies even after thinking about it, so what? Who says you can't eat cookies and still eat healthy? It's eating the whole bag while watching television that's really the issue – isn't it?

But what Hank really needs to learn is how to teach her to think, to evaluate and to decide. These are skills that aren't often taught to people with disabilities. And in teaching the skill is the transition of power, from him to her. Ah, that's the job isn't it? She should be learning to rely on herself. He needs to see that the "process" of making a decision is as important as the "decision" itself.

What happened here is that Sue learned only that staff have power and she has to do what they say. Like, that's news to her.

Finally, we need to look at how we behave in public when we're with people with disabilities. Are we teaching the public how to disrespect folks with cognitive disabilities?? Are we teaching the public that people with developmental disabilities don't have the same feelings and sensibilities as other people? Is this dangerous? I think so.

> *The higher a monkey climbs the more you see of its ass.*
>
> General Joseph Stilwell

Step Three: Detection through Introspection

Challenge yourself to grow. Challenge yourself to take every opportunity you can to really look at yourself and honestly evaluate how you are affecting other people. Challenge yourself to stay open to the fact that, even with the best intentions, you make mistakes.

North American society has become addicted to "victimology." I can be an absolute jerk and then forgive myself because my inner child wasn't nurtured enough when I was a kid. Well, excuse me but most inner children need a good spanking – and I don't agree with spanking. (Well, excepting the monkey who needs to be spanked routinely in my experience.)

I like being an adult, even though it means that I need to hold myself responsible for the things I've done. Without question I have regrets. I have done hurtful things. I have said hateful things. Whoever said that "Confession is good for the soul" was crazy. It hurts.

> *None who have been free can understand the terrible fascinating power of the hope of freedom on those who are not free.*
>
> Pearl S. Buck

Really hurts. But it also heals. When I look back at some of the things I've done to people with disabilities – from electric shock programmes to having parents make love contingent on appropriate behaviour – I grieve. But it would be a greater grief to not see error in the first place.

Someone once described a long-time staff person thusly, "He says he has twenty years of experience, but in reality he has had one year of experience repeated twenty times over." That is one of the saddest remarks I have ever heard about someone in this field.

I overheard an even sadder remark just last week when a woman with a disability was asked what her dream was and she said, "To do appropriate behaviours." I almost wept. I wanted to rant and rage against the staff who served her and the system in which she lived, but scant years ago one of my clients might well have responded, "To do what I'm told." While this woman wasn't someone I'd served, or even met, it still led me to the same place.

Quiet Thought.

Contemplation.

Evaluation.

Whatever you call it. Spend time looking at who you are and hold yourself accountable for what you've done and work to change.

> *Freedom is never voluntarily given by the oppressor; it must be demanded by the oppressed.*
>
> Martin Luther King, Jr.

Power Play (3)

The scene: A shadowy figure sitting quietly, and fully clothed, on a toilet (seat cover down).

The action: Noise is coming into the room from outside. While it can't quite be heard, it's clear someone is upset.

The player:

Kelly: a twenty-something staff working in a semi-independent living programme.

The dialogue:

Kelly: *speaking softly to self* OK, I blew that little interaction wide apart. I should know better. She doesn't like surprises and I dropped that "doctor's visit" on her kinda out of the blue. When she yelled at me, I know, I know I should have been just quiet for a second. I never should have tried to get her to sit down to talk. For me that seemed reasonable, for her it seemed forceful. When will I learn just to back off sometimes, gosh, I've always got to be the teacher, always got to know what to say. No wonder she thinks I'm

> *A life spent in making mistakes is not only more honourable but more useful than a life spent doing nothing.*
>
> George Bernard Shaw

just a control freak. Now, Kelly, remember this for next time. You are doing so much better. I don't think I've had to sit on this toilet and do this even once this past month. Now, how to get out there and solve this. Let's plan. Apologize – and sound like you mean it Kelly. Tell her you'll cancel the appointment if she wants. Get her to think it through. She knows that she needs to see the doctor, she requested the appointment anyways. But she needs to decide this. Help her to just pause and think ... but do it when she's ready. Oh, and Kelly, when you get really pissed off inside – remember you are being paid to do this. Take a breath and let's go.

First thing you need to realize. Kelly is pretty good at the job and is getting better by the minute. Kelly knows that Time Out ain't just for *them*!! Second thing you need to ask yourself is "What kind of great person is Kelly going to be in a year or two?" Ain't Kelly on a cool trip? Really cool?

Skill Number One: Scene Not Heard

Take a look at the following four scenes and see if you can find the common element that links them together.

1)   I am speaking at a conference at which several self advocate are attending along with their "care providers"

and while my audience is doing some small group activity, I take the opportunity to hit the bathroom. Because of my size I always use the stall for folks with disabilities. I fling the unlocked door open and see a man with cerebral palsy laying back in his wheel chair. His pants are unzipped and his erection is poking through the opening. A woman with Down Syndrome kneels in front of him giving him a blow job. They are startled at my entry and look at me with fear. I say, "Sorry," and leave.

> *Power never takes a back step – only in the face of more power.*
>
> Malcolm X

2) A woman, told that she can't watch her soap opera tapes until she has done her exercise programme picks up a table lamp and smashes it down on the floor. The staff grabs her and drags her kicking and screaming to her room for a Time Out. Once in the room the woman yells obscenities at the staff suggesting that the staff has an odd sexual attraction to neighbourhood dogs. The staff opens the medication chest and gets a PRN. Two other staff join in and the three of them pin the woman to the bed and hold her down until the needle punctures her ass-cheek and the medication is delivered. She is told to spend the rest of the night in her room calming down. It is four-thirty in the afternoon.

3) Two men with disabilities decide that they want to join with two others that they know from the local "Friendship Club." This is fine with the staff until they find out that the other two also have disabilities. The staff

> *The Bill of Rights ... is past due.*
>
> Peter McWilliams

decides that this kind of "self-segregation" is not appropriate and the men are told that they are not allowed to go out with the folks from the "Friendship Club" and are further informed that these guys aren't "Friends." They are encouraged to make "real friends" with non-disabled people. One of the two men bursts into tears and says, "but we promised," the other with a history of self-injury begins to slap himself and is eventually restrained. The staff later decide, at a meeting attended only by staff, that these two men are being "segregated" at the Friendship Club and are no longer permitted to attend.

4) During a routine room check a young staff discovers that a 56 year old woman had hidden a Barbie Doll deep in the back of her bottom drawer. This discovery led to the discovery of a stash of Barbie Fashion stuffed into an old pair of socks. Given that these things are age inappropriate for a woman of advanced age, the toys were gathered and dumped. When the woman returned home, she was confronted about the toys and told that she wouldn't be allowed to go shopping at the local mall which had a large toy store in it. She would have to go to the outdoor mall downtown – the staff had checked and there were no toy stores in that area. The woman reacted by grabbing the staff's glasses and breaking them. She was written up and punished for her aggression.

So what connects these four things? Well, given this is a

book about power and its misuse, you should be up to a pretty good guess. That's right each of these is a description of power gone mad. Let's just briefly look at each. If you are really

> *Distrust all men in whom the impulse to punish is powerful.*
>
> Friedrich Nietzsche

into this, and I hope you are, highlight each part of each sentence in the paragraphs that describe some kind of abuse of power. But, let's look at the basics.

Scene One: These two people were in the bathroom because the powers-that-be had determined that sex was not to be allowed in the group home. They were forced to find the most private place that they could. *This is not inappropriate behaviour, this is an understandable reaction to an inappropriate system.*

Scene Two: This woman is reacting to being told that her choice isn't valid. Not only is her choice invalid, her input into the discussion is dismissed. She is enraged that she can't be heard when calmly discussing things, so she decides that smashing lamps and loud vulgar words might get the message across. *This is not inappropriate behaviour, this is an understandable reaction to an inappropriate approach.*

Scene Three: These two men are reacting to the hidden message that they are "not good enough" for friendship. By saying that they can't have friends with others who have disabilities they are being spoon-fed self hatred. Tears and self-injury are a natural reaction to being told that you

> *The shepherd always tries to persuade the sheep that their interests and his own are the same.*
>
> Georges Stendal

are not a "real" person. *This is not inappropriate behaviour this is an understandable reaction to an inappropriate message.*

Scene Four: Smashing glasses seems hardly strong enough response considering her stuff was gone through, stolen and then destroyed. Further, punishment is banishment. Who are the staff – royalty? *This is not inappropriate behaviour it is an understandable reaction to an inappropriate philosophy.*

Learn that old axiom: For every action there is an equal and opposite reaction. People with disabilities, like others who are oppressed, will fight back. However, the person with a disability will probably not invite you to a political or philosophical discourse on service provision. No, they are most likely to want to rip your tongue out and then beat you around the head with it. Crude polemics but it gets its point across.

Being able to entertain the idea that someone's behaviour might be a reaction to oppression leads to a greater ability to Reflect. Remember, though, not every problem that a person with a disability experiences will be because of power, or lack thereof, but enough that it's worth considering each and every time. Playing connect the dots with a series of problem behaviours in a group home may lead you to realize that the folks with disabilities are staging an insurrection.

Skill Two: Power Sharing

> *Example is not the main thing in influencing others. It is the only thing.*
>
> Albert Schweitzer

If you think that the solution is for you to speak up to change the system ... you're on the right track but heading in the wrong direction. Go back and re-read Step Three ... there's another think coming. Doing something for people with disabilities isn't doing them a favour. If there is oppression, cool – recognize it, then help a person with a disability figure out what to do about it. And there are always options from making a change in how chores are assigned, how staff make suggestions, when dinner is served ... to forming a self advocacy group to learn how to better speak out.

Self advocacy happens in groups but it also happens between you and someone with a disability. Help them learn to respectfully, but firmly, disagree with you. Listen to their arguments and give the "extra five minutes needed" for them to formulate their point of view. Try to turn off the debate mode and remember they are talking to you about *their life* and it's only *your job*. Be wrong sometimes, for heaven's sake. Back down from a point of view, a command given, an intrusion made – that one moment will teach more than anything else you have done that day. Self advocates need to practice standing up for their rights, why not you ... you're a safe person who works with the downtrodden and you've got your power issues under control. Right?

> *God gave me an extra chromosome. It's an odd gift, but it's a gift.*
>
> Astra Milberg

Come to see disability as a unique perspective on the world. Instead of always trying to teach them how to see the world in the "right" way, remember that there is no "right" way to see the world. "Retarded people don't do it our way." Toby taught me that and I think it's kind of a cool thing to realize. Instead of narrowing their vision to fit ours, maybe we need to broaden our vision to include theirs.

I watched a woman with a disability get on a witness stand and tell how some had violated her trust. She used power well – and her action protected other people with disabilities from harm.

I watched a man with a disability stand up to a bully who kept calling him a "Retard" and an "Idiot." He used power well – and his action silenced a bigot.

I watched a young staff stand and quietly support an older woman as she told her staff that she wanted her Barbie doll back and her privacy respected. These two women used power well – one supported, one relied, both changed.

> *Let my people go.*
>
> Moses

Justice is Just Us. Just us, all of us, in it together.